LEVEL
1

Hippos

Maya Myers

NATIONAL
GEOGRAPHIC

Washington, D.C.

For Buzz and Helen, my favorite hippo-loving pair —M. M.

Published by National Geographic Partners, LLC, Washington, DC 20036.

Designed by Gustavo Tello

The author and publisher gratefully acknowledge the literacy review of this book by Mariam Jean Dreher, professor emerita of reading education, University of Maryland, College Park, and content review by animal and education experts at Disney's Animals, Science and Environment.

Library of Congress Cataloging-in-Publication Data

Names: Myers, Maya, author.
Title: Hippos / Maya Myers.
Description: Washington, DC : National Geographic Kids, 2024. | Series: National geographic readers | Audience: Ages 4-6 | Audience: Grades K-1
Identifiers: LCCN 2023030645 (print) | LCCN 2023030646 (ebook) | ISBN 9781426377020 (paperback) | ISBN 9781426377068 (library binding) | ISBN 9781426377051 (ebook) | ISBN 9781426377044 (ebook other)
Subjects: LCSH: Hippopotamidae--Juvenile literature.
Classification: LCC QL737.U57 M94 2024 (print) | LCC QL737.U57 (ebook) | DDC 599.63/5--dc23/eng/20230908
LC record available at https://lccn.loc.gov/2023030645
LC ebook record available at https://lccn.loc.gov/2023030646

Photo Credits

Cover, Beverly Joubert/National Geographic Image Collection; (VOCAB THROUGHOUT), robuart/Shutterstock; (HEADER THROUGHOUT), Aleks2T22/Shutterstock; Zhenyakot/Shutterstock; 1, Henk Bogaard/Adobe Stock; 3, Eric Isselée/Adobe Stock; 4-5, Uryadnikov Sergey/Adobe Stock; 6-7, Lost Horizon Images/Getty Images; 7 (LO), Radek Borovka/Shutterstock; 8, laverrue was here/Getty Images; 9, Peter Scoones/Nature Picture Library; 10, paula/Adobe Stock; 11 (UP), Rostislav Stach/Shutterstock; 11 (LO), snaptitude/Adobe Stock; 12, Jami Tarris/Minden Pictures; 13 (UP), photocech/Adobe Stock; 13 (LO), kstipek/Adobe Stock; 14-15, Bavorndej/Adobe Stock; 15 (UP), EcoView/Adobe Stock; 16, Uryadnikov Sergey/Adobe Stock; 17, Jurgens/Adobe Stock; 18 (UP), tk/Adobe Stock; 18 (LO LE), Stu Porter/Shutterstock; 18 (LO RT), ZSSD/Minden Pictures; 19 (UP), Eric Isselée/Shutterstock; 19 (CTR), Videologia/Getty Images; 19 (LO), jtplatt/Adobe Stock; 20, Jane Rix/Shutterstock; 21 (LO), Faas/Adobe Stock; 21 (UP), mr_riyazin/Shutterstock; 22-23, Christopher Scott/Alamy Stock Photo; 22 (LO), sathit savettanant/Shutterstock; 24, ZSSD/Minden Pictures; 24-25, ZSSD/Minden Pictures; 26, Joe Petersburger/Getty Images; 27, stu-porter/Adobe Stock; 28, Image Source Trading Ltd/Shutterstock; 29, JossK/Adobe Stock; 30 (LE), chbaum/Adobe Stock; 30 (RT), MyImages - Micha/Shutterstock; 31 (UP LE), flytoskyft11/Adobe Stock; 31 (UP RT), Eric Baccega/Nature Picture Library; 31 (LO LE), Steve Gettle/Minden Pictures; 31 (LO RT), Tami Freed/Shutterstock; 32 (UP), Isselée/Dreamstime; 32 (CTR LE), ChiccoDodiFC/Adobe Stock; 32 (CTR RT), Mint Images/Getty Images; 32 (LO LE), faruk/Adobe Stock; 32, Eric Baccega/Nature Picture Library

Printed in the United States of America
24/WOR/1

Contents

Howdy, Hippos!

Say hello to the hippopotamus! It's one of the largest mammals on land. Even though they are land animals, hippos love water!

Hanging With Hippos!

Hippos make their homes in lakes, rivers, and swamps in Africa. These giant animals spend most of their day in the water. They must be good swimmers, right?

Nope! Hippos can't swim.
Their bodies are too dense to float.
They walk or run along the bottom.

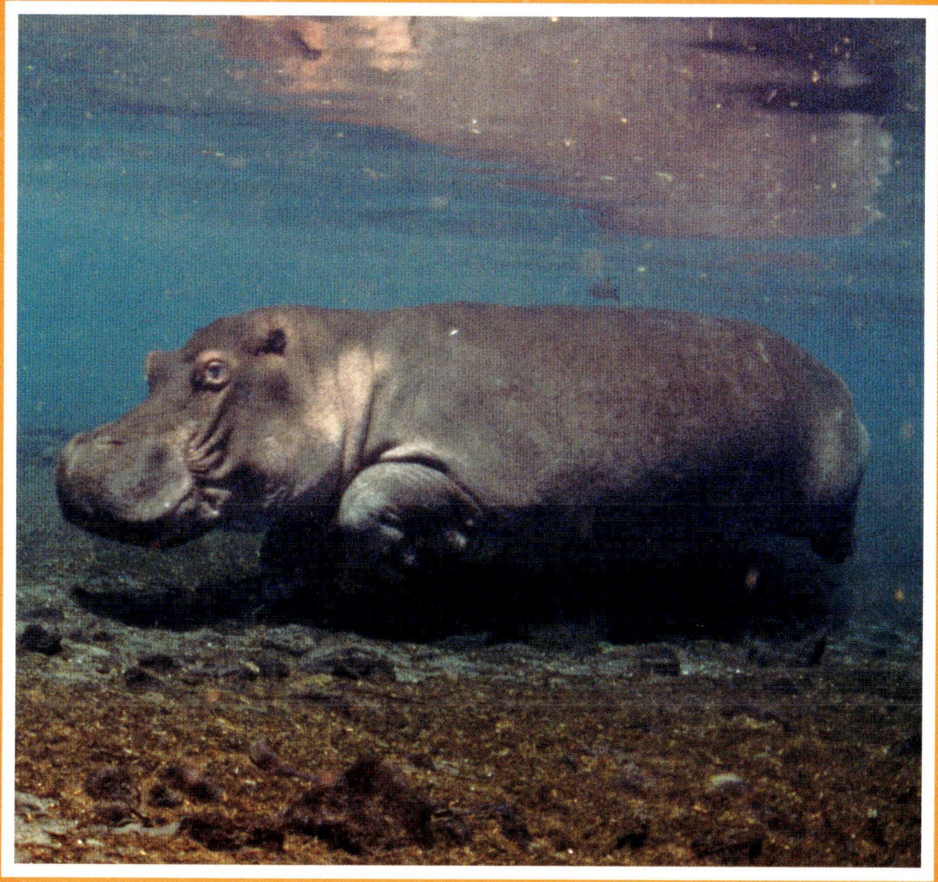

They push off from rocks and glide through the water.

Hippo Hint

DENSE: Very heavy for its size

Hippo Herds

Hippos live together in groups called herds. A herd might have 10 hippos or as many as 100. A herd includes young hippos and adult males and females.

Hippos protect their herd and their territory. Male hippos fight to decide which one is in charge.

Stay away! It's important to keep a safe distance from hippo habitats.

Hippo Hint

TERRITORY: The place a group of animals lives in and protects

Huge Hippos

Adult hippos can weigh about 2,500 to 5,500 pounds—as much as a car! But they are shorter than most adult humans. Let's take a look at this big body!

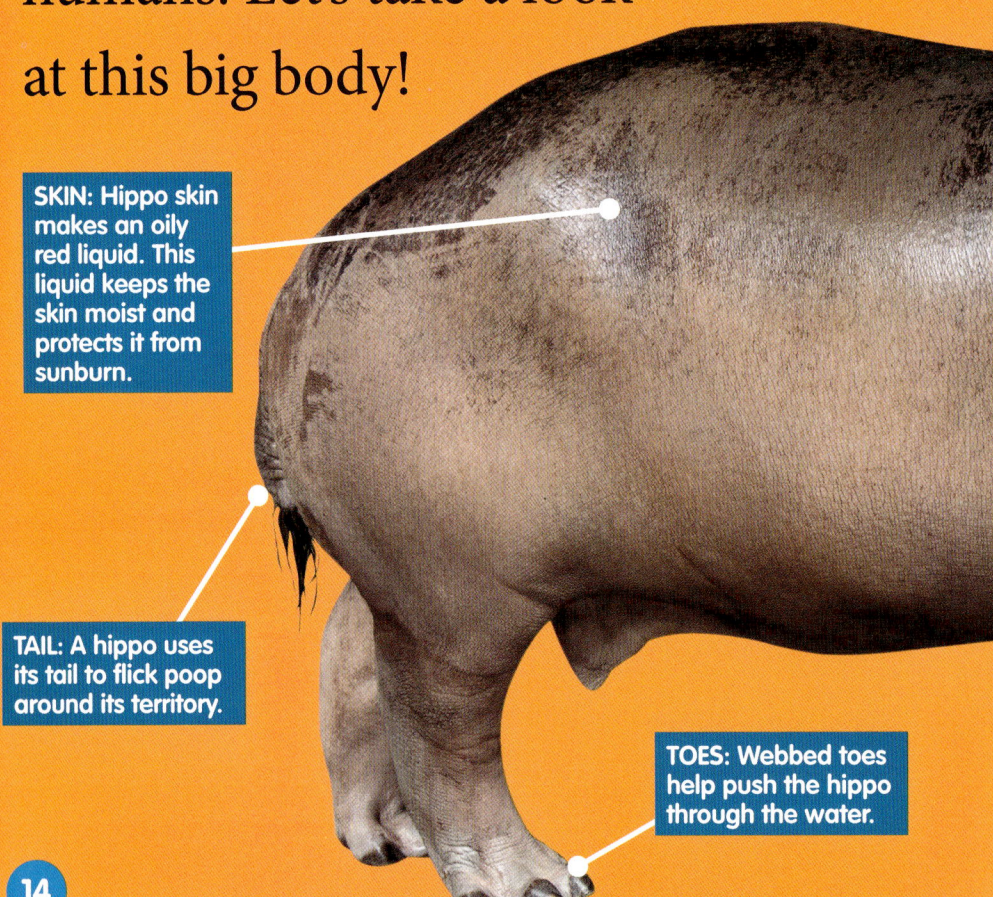

SKIN: Hippo skin makes an oily red liquid. This liquid keeps the skin moist and protects it from sunburn.

TAIL: A hippo uses its tail to flick poop around its territory.

TOES: Webbed toes help push the hippo through the water.

NOSTRILS: A hippo can breathe even when most of its body is underwater because its nostrils sit high on its head.

EYES: Eyes on top of its head help a hippo see all around. Clear membranes over the eyes make natural goggles so a hippo can see underwater.

EARS: A hippo's ears (and nostrils) can close up tight to keep water out.

HAIR: Thick, stiff hairs are found only on a hippo's head and tail.

Hippo Hint

WEBBED: Connected by skin

Honking Hippos!

Wheeze honk!

A hippo bellows to warn the herd of danger. Other hippos join in.

Hippos communicate with many sounds. They recognize the voices in their herd. A hippo honk can be as loud as thunder.

6 FUN FACTS
About Hippos

1

Pygmy hippos are small—about the **size of** a **farm pig.** They **live only** in **rainforests** in West Africa.

2 A hippo can **open its mouth** so **wide** that **its jaws** almost **form a straight line.**

Hippo skin can dry out and **burn** in the sun. That's **one reason** hippos **spend** so much time **in the water.**

3

4 Some **hippo teeth** can be **20 inches long.** That's about as **long** as your **whole arm!**

If a **male hippo hears a strange** hippo honking, he **flings poop around with his tail.** This tells the **stranger** to **stay away!** **5**

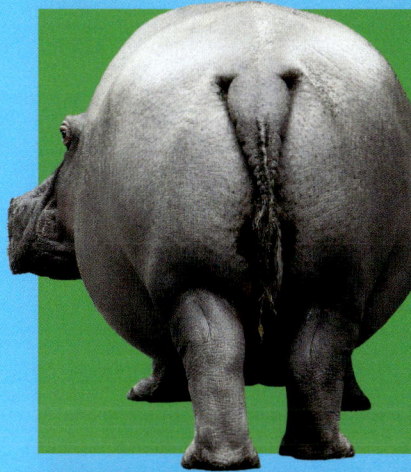

Hippos can **run** about **20 miles an hour**—faster than most people can **ride a bike!**

6

Hippo Huddles

Hippos sleep a lot during the day. Some sleep huddled in groups on the shore. Some sleep underwater.

Q What did the hippo get at the gym?

A Hippopotamuscles!

They hold their breath for up to five minutes, then come up for air without waking up.

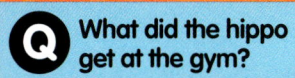

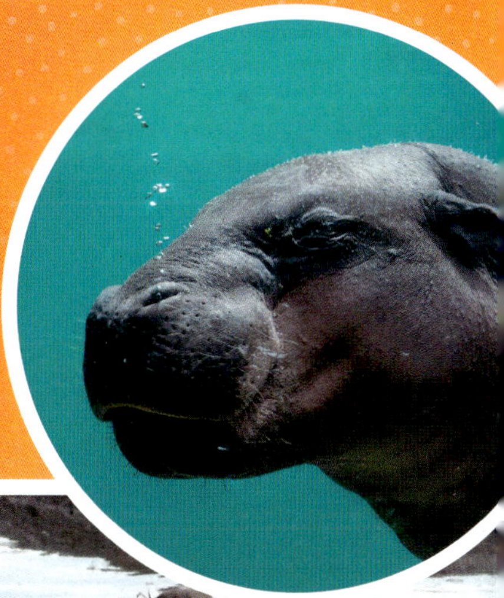

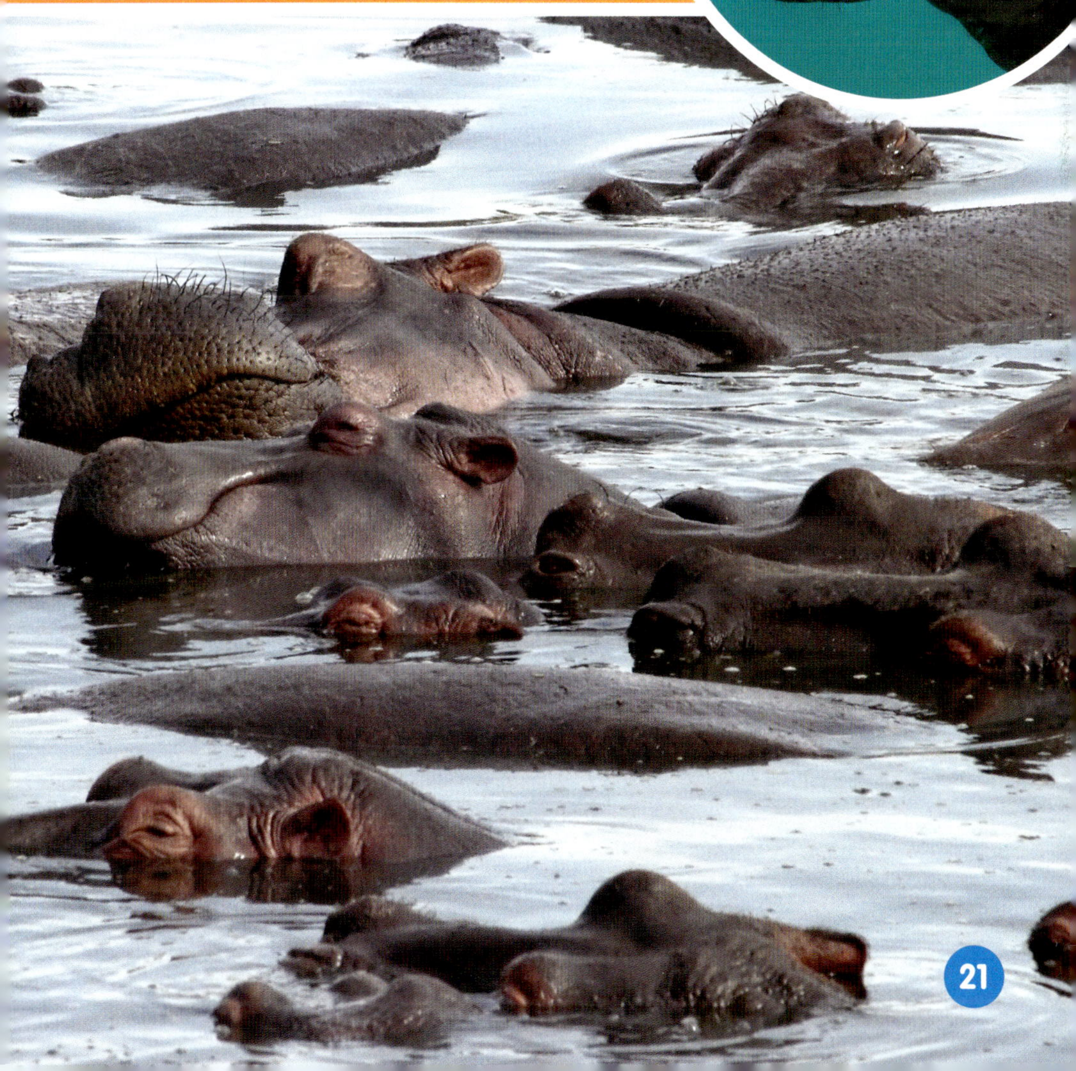

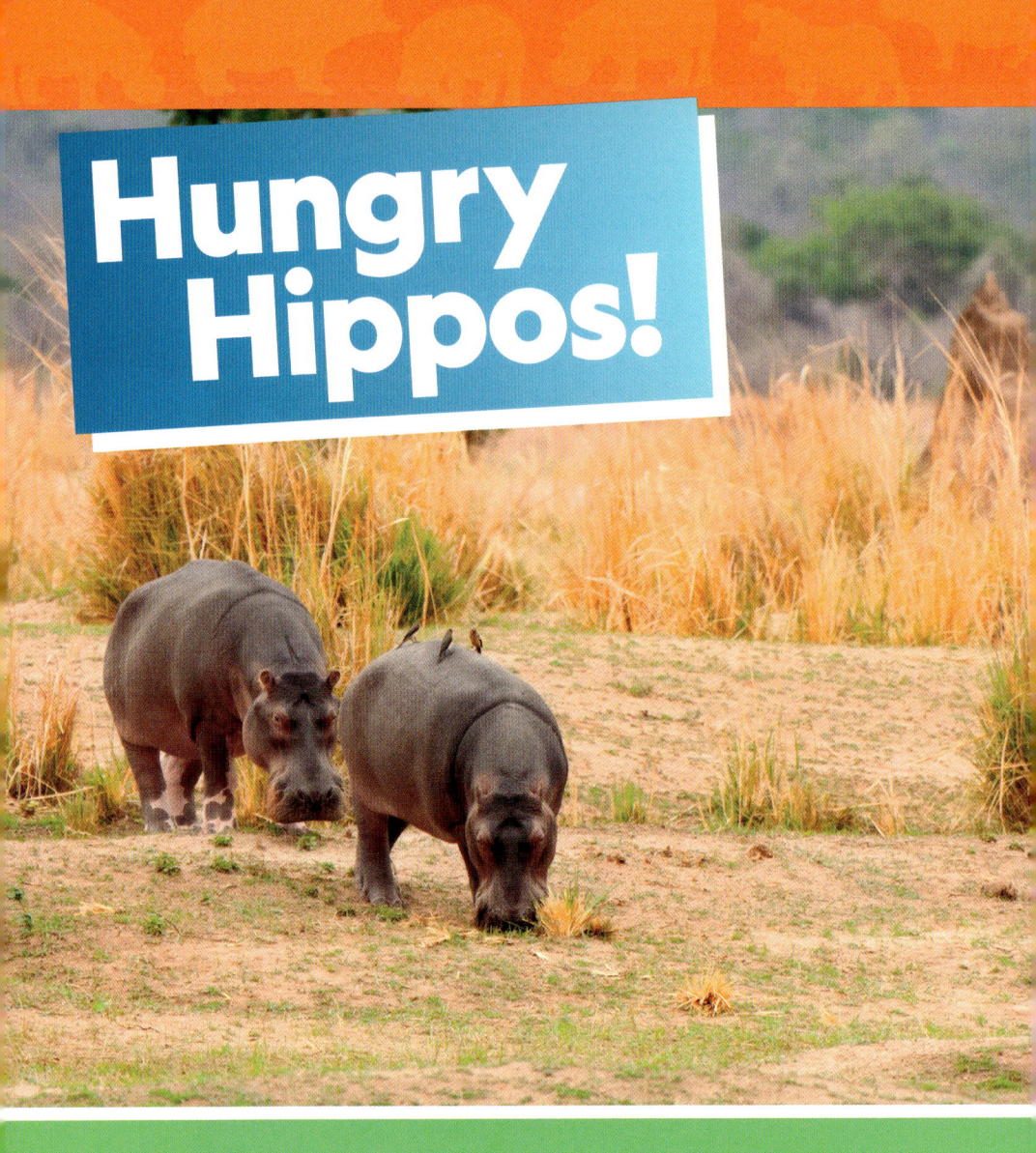

Hungry Hippos!

When the sun sets, it's time to eat. Hippos leave the water. They walk along paths to grassy areas.

They eat around 90 pounds of food every night! They munch on grass, fruit, and other plants.

Hippo Babies

A hippo mother moves away from her herd. It's time for her baby to be born. The calf is born underwater. The mother pushes it out of the water to breathe.

The mother and her baby stay in the water most of the time. The calf can even drink milk underwater. Its ears and nostrils close to keep water out while it drinks.

Around two weeks later, mother and calf return to the herd. The herd helps keep the calf safe from predators.

How Hippos Help

A hippo's poop keeps its ecosystem healthy! The poop contains important nutrients that help animals and water plants live and grow.

Q What happens when a hippo gets in your house?

A A hippopotamess!

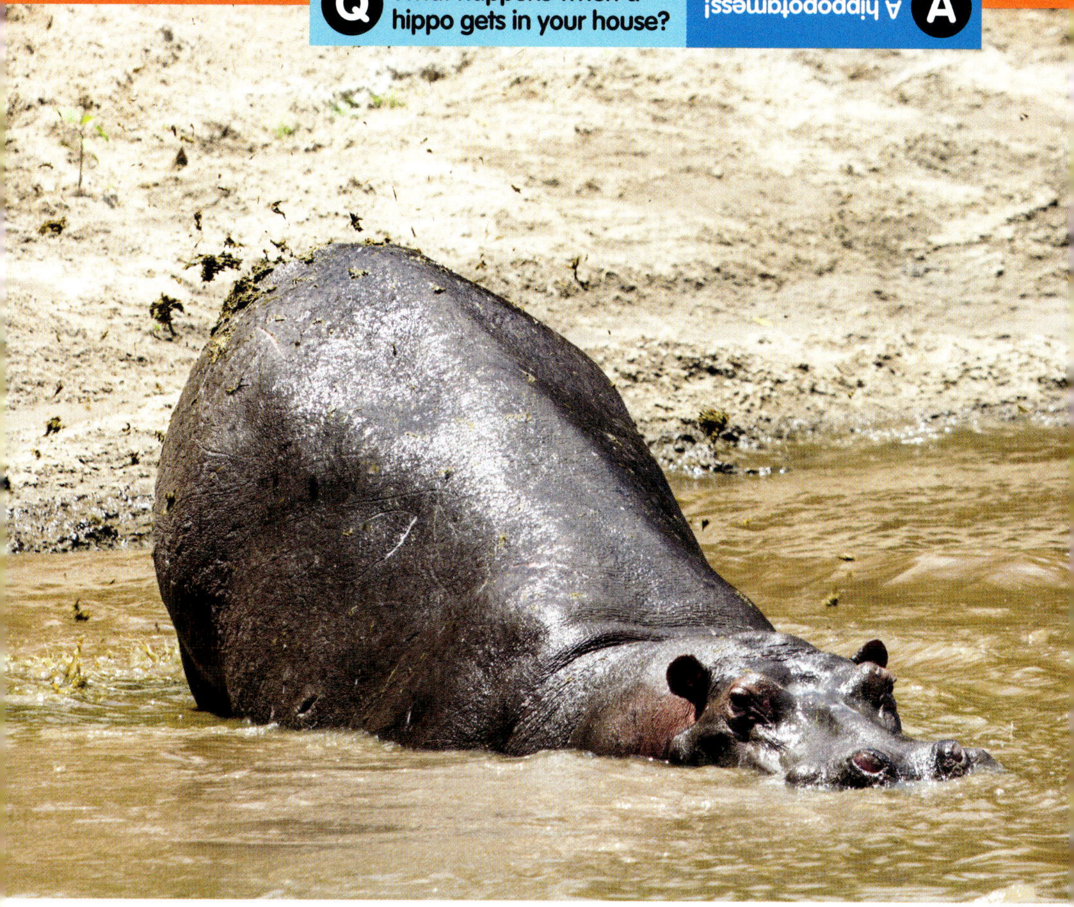

When hippos eat grass on land and then go to the river to poop, they move nutrients from the land into the water.

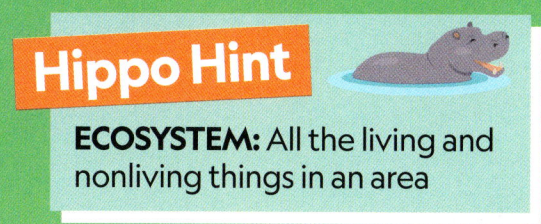

Hippo Hint

ECOSYSTEM: All the living and nonliving things in an area

What in the World?

These pictures are close-up views of hippos. Use the hints to help figure out what's in the pictures. Answers are on page 31.

1

HINT: These sit above the water to help hippos breathe.

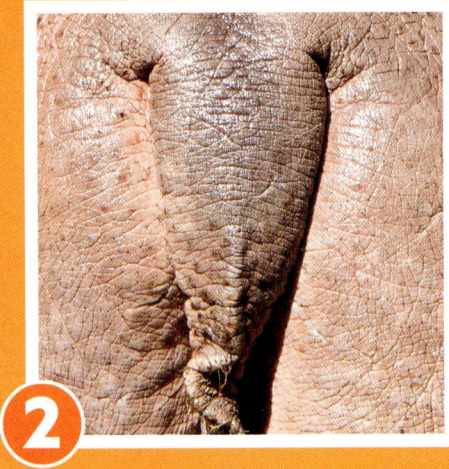

2

HINT: A male uses this to spread poop.

Word Bank

hair	toes	nostrils	ear	skin	tail

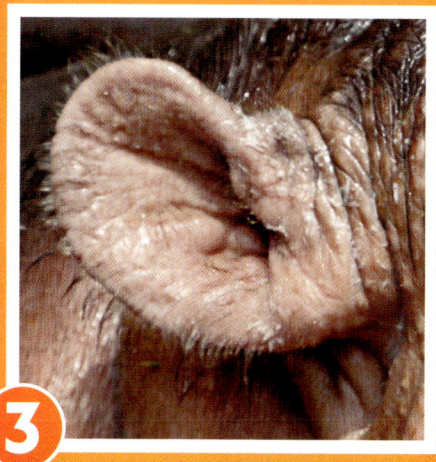

3

HINT: This can close to keep water out.

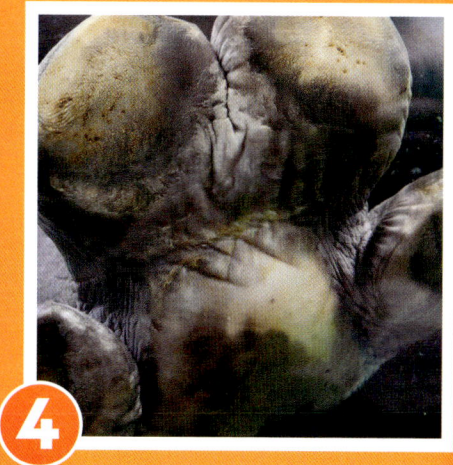

4

HINT: These are webbed to help push through water.

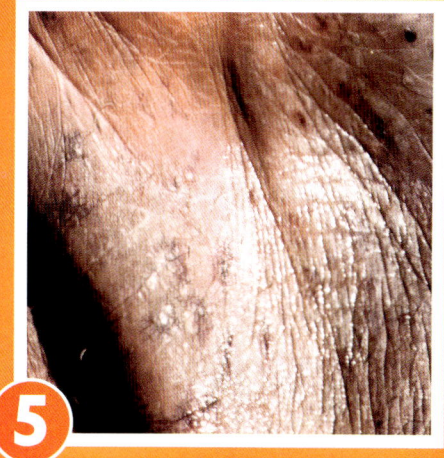

5

HINT: An oily liquid protects this from the sun.

6

HINT: This is only found on the head and the tail.

Answers: 1. nostrils, 2. tail, 3. ear, 4. toes, 5. skin, 6. hair

GLOSSARY

DENSE: Very heavy for its size

ECOSYSTEM: All the living and nonliving things in an area

TERRITORY: The place a group of animals lives in and protects

WEBBED: Connected by skin